How to Study Physics

This edition published in 2020 by Circlesquare Projections

Copyright© by Circlesquare Projections

ISBN 978-0-578-68406-2

Circlesquare Projections Publishing company

Pacoima, Los Angeles, CA, 91331

How to Study Physics

Jose Valladares

Circlesquare Projections Publishing Company

Contents

◆◆◆◆◆◆◆◆

Scope of this book .... *iii*

Preface .... *iv*

1 Why Study Physics? *5*

2 Developing the Study Habit *10*

3 How to Take Notes *15*

4 Mathematics in Physics *17*

5 Physics and Measurements *28*

6 How to Work Physics Problems *31*

7 Physics is so Complicated *42*

8 Experiments in Laboratory *46*

9 How to Take Examinations *49*

Appendix *53*

Bibliography *57*

HOW TO STUDY PHYSICS

Scope of this book

This book is written for students. It is intended to be used as a guide on how to study physics. My primary aim is to prepare the student on how to study physics for this reason I have presented topics in logical order to make it easy for a student to understand, emphasizing on mathematics and measurements in physics. We will not study in detail all the concepts of physics. The type of questions we will discuss are not mentioned in most physics textbooks. The questions we will discuss relate to your personal life, and to prepare you to succeed in a physics course.

Preface

For a student of physics is important, and necessary to ask questions. It is obvious that the real purpose of taking physics is not just to memorize concepts, and principles, but to train your mind to think clearly, and logically. If you don't ask questions now, early in the chapters, to train your mind to practice simple problems, later on more difficult problems won't be solved effectively.

You will remember the answers to your own questions the most. Learn to ask the right questions, and also ask yourself questions to end any doubts regarding physics.

Some questions you can ask yourself regarding a principle or a concept

- What is velocity?
- Why is it so?
- What is the equation?
- What is a typical problem concerning it?
- Do I know what to do with it? Do I really understand it?
- Why is it important to know it?
- Does it tie in with other ideas in physics?

If you ask yourself these types of questions, and you have precise answers then you will do well in physics. It is essential to organize all your thoughts, and remember your answers.

CHAPTER 1

WHY STUDY PHYSICS?

What is physics? Physics is the search for truth, a generalization which applies to all sciences. In physics we ask "how fast can you fall?", "why do we sink?", or "what makes materials stretch?", If we can find answers to these questions, we make progress as individuals, our psyche expands to better understand the world around us. Tremendous possibilities opened up to promote the progress of mankind, and to discover useful things to improve the world we live in.

Physics is all about us. Physics is a basic science. It comes into play in the field of Medicine, Biology, Architecture, Technology, Earth and Environmental Sciences. Physics answers many questions as to "how?" or "why?". Yet, not everyone has an opportunity to study physics at school or college. This, then, is the excuse for a book on how to study physics, a student can ask himself, "Can I obtain a basic knowledge of Physics without long hours reading the textbook or long hours in the laboratory?". The answer to this question is "Yes". It is possible to understand, and obtain a working knowledge of Physics, by diligently, progressively, and patiently study the subject carefully. I assume also that a student of science will not object to a word of advice, and to some suggestions as to how to study physics.

The main purpose of this book is to improve your ability to solve physics problems, as the main objective of an education is to train people to think clearly about what they want to achieve in life. This book is primarily intended for students who are going to college, who are currently taking elementary physics or engineering course, who already know what they want in life, and have an adequate idea of finding the best methods of learning a subject; in this case their chief activity will be studying. Yet, it is not required to attend college to study physics, this book can also be a guide for those students who don't have an opportunity to study physics at college or school.

The primary requirement to learning a material in life is mental attitude. You must have a desire to learn. The first step to learning is to say "I don't know anything", as Socrates once said that it takes great wisdom to accept one owns ignorance. If you are firmly convinced that you already know everything then you will not have a desire to learn, and this book will do little good. One has to be honest with oneself, a student is honest with himself when he sits down, opens the first page of the book, and begins to learn. Knowledge begins when one honestly accepts that one doesn't know.

We all have different reasons to want to learn physics, and different mental abilities to acquire its definition and concepts. There are many different types of intelligence, and you should never doubt your intelligence to learn physics. You are capable to achieve anything in life that you set your mind to do. If you are able to say "I'm not capable", then for sure you are not capable, because you have placed a limit in your mind. We all have intelligence. It is a gift. It is a gift we all poses. It is built into our DNA. We must never placed any limit in our intelligence.

It is certain that not all students learn effectively in the same way. As you study physics, try out different ways to learn, and slowly develop a system of study that perfectly suits you. Later on, you can apply this system to other science subjects. Honestly, learning physics takes work, it is time consuming to read and solve word problems. There are no shortcuts to learning physics, this book can only guide you so that you may work more effectively.

Follow suggestions covered in this book, and it will help you get a better grade in your course. It's entirely up to you to decide, if you want to devote most of your time to get an A in physics, but then if you do that you may fail other courses from having to spent all of your time in physics.

This book will help you use time more effectively. I will cover specific examples, and a summary of the main idea instructors I want you to pay careful attention to.

Physics for doctors

Doctors of medicine, and science departments understand the importance of physics in Medicine, and these are important application of physics: x-rays, nuclear medicine, clinical PET scanning, magnetic resonance spectroscopy, magnetoencephalography, high intensity focused ultrasound with MRI, radiotherapy treatment, and interventional MRI. Medical physicists play an important part in society, for example patient diagnosis, and disease treatment. An otolaryngologist needs to understand mechanism of hearing to diagnose and treat his patients.

An Ophthalmologists needs to know about problems of physiological optics and vision. A physical therapist needs to understand center of gravity, and the principles of forces within the human body. A physiologist needs to know about nerve induction, and nerve contractions. All of these are in the domain of physics, and important applications of physics to Medicine.

Physics for engineers

Engineers, and scientists will learn how to apply principles of physics to solve problems, and answer many other questions. You will encounter the standard topics of mechanics, sound, light, heat, electricity, magnetism, atomic physics, and nuclear physics. You will need mathematical knowledge of algebra, geometry, and trigonometry. Engineers carry a heavy program of studies, honestly, there won't be enough time while you are in college to learn everything in detail. Therefore, it is essential to learn how to teach yourself, and decipher all the facts you will need for exams.

The general student

Physics is the most basic of the sciences. It is usually divided into motion, force, and energy. These fields are motion, fluids, heat, sound, light, electricity, atomic structure, nuclear physics, astrophysics elementary particles, and magnetism. Physics deals with the behavior of particles, and the structure of matter. You don't have to be a genius or a research scientist in medicine to able to use physics in daily task. It is essential to understand basic concepts, principles, and to apply physics in daily tasks of life.

If you are interested to study physics to learn how everything works, examples such as how an electric motor works, or how a telephone works, or the size of an atom of hydrogen, or how we discover these things, then it is necessary for us all to know these things.

It is our duty as intelligent beings to understand the basic principles of physics. A person who discusses atomic energy, gravity, and motion understands that this knowledge is not only for scientists, and engineers. The responsability on how to solve world problems shouldn't be for only scientists, and engineers, but since physics is about us; every individual plays a role on world problems.

The Physicist

A person who wants to become a physicist may be expected to know the following:

1.   The concepts and principles of physics

        It is essential for a student to understand concepts and principles before attempting to solve a problem.

2.   How to ask yourself questions (methods of perceiving problems)

        Physics is about perception. The way you perceive a problem is by creating a model in your mind, apply your knowledge to your model, and then ask yourself questions whether the principles and concepts of physics make sense. Before you perform a calculation it is important to get to the heart of the problem.

3. Perception(How to separate essential from the non-essential information of a problem)

A physicist develops mental intuition over time by solving many different types of problems. He learns different techniques to attacking problems. These techniques are acquired in the laboratory.

4. Laboratory

Your understanding of the material in the lectures is enhanced in the laboratory through a combination of studying before going to class, discussing the materials with students, and instructors, and ability to solve problems.

It is obvious for a physicist to have a clear understanding of concepts and principles of physics. Perception, and laboratory knowledge are equally important, and can be acquired by practicing on solving problems.

The methods and techniques are important, and it is essential to train your thinking through practice on simple problems, so it is important that you understand, be able to apply various concepts and theories discussed in class. The simple problems will build self-confidence, so that later on more difficult problems, and situations can be approached more effectively.

Solve physics problems

Solve as many problems as you can in the first two chapters, keep in mind that you will not absorb the full meaning of the terms after one reading, and several readings of the chapters, and notes may be necessary. Ask yourself questions while reading, and don't be afraid to ask questions if you have any doubts.

CHAPTER 2

DEVELOPING THE STUDY HABIT

Positive mental attitude

The most important requirements to learn any subject, either mathematics or physics, is positive mental attitude, and a driving desire to learn. Physics is not an easy subject, this is the reason why is recommended to maintain a positive attitude towards the subject. If the course is required as part of a curriculum of professional training, then the course is necessary. The main objective of physics is to train people to think clearly, to develop organized thinking, and to demonstrate to the instructor that you can understand science. Physics teaches order. The material and problems in physics teach students how to think clearly, and clarity of thought is driven by order. Everything in the universe has order. Order and disorder can't coexist together. If there is disorder in your mind, physics requires you to have order, in a sense that to understand concepts of physics requires you to have order in your mind. Order is like the blood for clarity of thought, when you understand the material you can easily apply the knowledge on paper, and explain the concepts to someone else with confidence. Physics is not about intelligence is about order. Physics becomes easy to a student who is organized.

As a student of science, physics can be learned by seeing, hearing, reading, writing, and talking. You will be in a class with other students who also want to learn physics, get interested in the subject by talking things over with fellow students, and perhaps you will develop more enthusiasm by learning physics with your friends.

Go to class on time. Be yourself. Be alert. Be ready to learn. Be cooperative with students and instructor. Make sure to read the syllabus for the course and follow the schedule set by your instructor. If you easily get distracted sit in first or second row, where you won't get distracted by fellow students, and you will be forced to pay attention. Make it a habit, and always plan ahead on each day before lecture to do a careful reading of the textbook before attending your class. Keep in mind that few people can absorb scientific material after one reading.

The instructor will repeat the material you have already covered ahead of time, and physics will make more sense. Take careful notes, because the instructor has a plan; he wants you to learn from his notes. Ask questions on ideas that required clarification.

Study physics daily, you can't study a chapter for one hour, then not study for the next 3 days, and expect to learn. It is important to create your own study schedule, set time aside for other classes, and devote a certain amount of time to study. I recommend to arrange a study schedule, read below for instructions, at least one hour per day.

You learn more physics by studying it for an hour a day than by studying it on weekends for 8 hours. It is much easier to learn physics from day to day than learning the entire chapter in one day. If you are taking multiple units, it is better to study one subject for an hour, and then shift to another subject after one hour. During study session of several hours, do an occasional relaxation. Don't get behind, and try not to force your mind to learn physics for hours. Experiment to find out which method suits you. Keep up with your work.

You should reduce memorization of physics to a minimum. Memorizing passages of a text, equations, and writing derivations on paper doesn't mean you understand physics. To master physics, or any subject, takes work, and that takes time. Your understanding of the material is a combination of laboratory experiments, discussing material with instructor or other students, and by going over the material on daily basis. In physics, it is essential to develop an ability to analyze problems, to think logically, and discriminate between essential and non-essential material. Study to understand the material, read carefully, and go slowly. Physics can't be read like a novel, or not even like a history lesson. Try to think of applications of the material as you read physics, and pause for a moment to think how equations apply to theories, or how formulas are derived.

While studying, keep personal worries off your mind. If you have emotional issues, get some good advice, think it over, and try not to force your mind to study while you have personal problems. You will not be able to learn, and anything you go through will be forgotten.

It is important to budget your time. Make out a study schedule, preferably on a daily basis, and stick to it for at least two weeks. Get enough sleep. Do regular moderate physical exercise, and some recreation. As a general rule, leave two hours of study per class.

Find yourself a quiet place to study, either at home, or at the library, with plenty of light, and desk space. Make sure your desk is free from distractions, that is away from tablets, phone, or computer. Avoid replying to text messages, one little distraction will disrupt your train of thought, and then it's not easy to get back on again. Study conscientiously. Sit with your back to the door, and reject all kinds of interruptions. It is important that you avoid the habit of not studying for days, a useful advice, not studying will lead to disastrous results.

Plan to study physics as soon after class as possible, while you still remember things that will probably be forgotten the following day. You will be able to solve problems after class while your mind is still fresh requiring less concentration. An occasional relaxation period of 15 minutes often is a help or getting something to eat before studying may help. You will not get much done if your force yourself to study with an empty stomach.

When you finish a chapter, try to solve odd-number problems, and the answers are given at the end of the textbook. The purpose to solving problems is to test your problem solving skills as you read through the textbook. After you have finish solving problems, think for a moment what you learned, and think out the main ideas. Write main ideas down or say them in your mind; keeping a record of your thoughts. On the day of the exam, your instructor might ask you to write down the principles in your own words.

Ask yourself questions to test your knowledge. You will learn more effectively when you put your knowledge to the test, by asking yourself what you learn about. Try to explain to a friend what you learn to find out whether you can answer questions, if you have doubts, then its time to do a second reading of the material.

You will find out that physics questions have simple yes or no answers, and the way to ask yourself questions is by over analyzing the material, or until all your doubts have been answered.

Some questions you can ask yourself when you have studied newton's gravitational law and newton's three laws of motion:

- Does a heavy object such as a rock drops at the same time from the same height as a light object(feather)?

- Does the speed observe of a moving object depend on the observer's speed?

- Does a particle exist or not?

You will find out some answers don't turn out as you first expect. In physics it is important to always ask yourself questions, if you can't find answers to your questions contact your instructor for help.

Study schedule

It is important to setup a study schedule, if you work full time, and study physics. Make sure to read the syllabus for the course before the semester, and start planning on a daily study schedule. Learn to budget your time. One idea I recommend to write down on study schedule is take a daily nap. Taking a nap before physics will help you pay attention better, and enhance your memory. If it's not possible to take a nap before physics at least take one before studying for physics.

On average per week you will have 48 hours to study. Most of time will be lost doing errands, replying to text messages, reading the news, and watching shows. It is entirely up to you whether you want to make good use of your time. It is important not to get behind, and always be on top of things. While studying physics you may be lost daydreaming, as it happens to physics students who try to find solutions to hard problems. If you have been thinking hard, and can't get the answers, take a 10 to 15 minutes break.

In order to help you save time I will list constants, and conversion factors. I will post this useful information on the back of the book.

CHAPTER 3

HOW TO TAKE NOTES

We all know how to take notes. We know we normally write down important ideas, or information we feel we might forget over time. Taking notes in physics lecture is quite different than taking good notes in History, or English class. One main difference is that most history or English classes are the representations of historical materials or fictional ideas. A student in history takes good notes to memorize the material. In physics you should reduce memorization of the material to a minimum. Whereas, in physics, lectures are primarily the explanation of principles, or concepts. These principles are illustrated by demonstrations and examples. An instructor will write down a formula on the board, and will demonstrate how the formula applies to a diagram on the board. Make sure to draw the diagram, don't omit symbols or arrows, as these arrows normally represent vectors or forces acting on objects. If you have read the chapter the previous day don't assume you already know the material, and assume what the instructor will show you will be pointless, it never hurts to repeat the material in your mind.

The same concepts, and principles explained in the chapter the instructor will also explain them in his or her own words. The main thing to do while taking notes is to write down the explanation. You will actually understand the explanations better if you spend some of your time studying before class.

For most lecturers the instructor will demonstrate two to four pages of notes per day. Some students will write nothing on paper except diagrams, while others take as much notes as possible to maximize their learning experience.

If the instructor goes on too fast, ask him or her to slow down, don't be afraid to ask; you may ask to repeat the problem or to clarify a concept.

Phycologists state that process of writing notes on paper contributes to the learning process, in addition to the verbal explanation helps you concentrate and remember the material better. After class discuss the lesson together with a fellow student, or make a statement regarding the lesson to your instructor to get more insight knowledge. Physics instructors are always willing, and ready to give insight information. The main purpose of taking good notes is to help you understand the material, and a double purpose of being a learning aid physically before an exam or quiz.

At night before going to sleep browse through your notes, as a way to review your day, and help you remember how to apply formulas.

CHAPTER 4

MATHEMATICS IN PHYSICS

Physics is not an easy subject, the reason why is not easy, it's not physics itself, sometimes it is the mathematics used that is the source of the difficulty. A student may imagine that physics is a difficult subject when actually may be his mathematical background which has been forgotten or too rusty to be useful. If you are worry or concern, then you need to review simple mathematical techniques, including algebra, geometry, and trigonometry. You will find this section useful to examine old topics or learn any new ones. These sections in mathematics are intended as a brief review of operations and methods. I have included practice problems as a way to refresh your abilities to solve math problems.

Elementary operations

I assume that you know how to add, subtract, multiply, and divide. Some long operations, like divisions, and multiplications I recommend that you use a calculator.

Exponents

The notation x^n, x is a quantity multiplying itself times the number of n. For example, $x^2 = x \cdot x$, and $x^3 = x \cdot x \cdot x$. The quantity n is called the exponent, or the power of x(the base). Some rules that will help you simplify

When two powers of x are multiplied, the exponents are added:

$$(x^m)(x^n) = x^{(m+n)}$$

When $n=0$ the power is defined to be 1

$$x^0 = 1$$

When two powers are divided

$$\frac{x^n}{x^m} = x^n x^{-m} = x^{n-m}$$;the exponents are subtracted

When a power is raised to another power,

$$(x^n)^m = x^{nm}$$;the exponents are multiplied

Practice problems:

$$x^6 x^0 =$$

$$\frac{x^9}{x^{1/3}} =$$

Scientific notation

In physics or in science in general many quantities are often very large or very small values.

For example, the speed of light is 300,000,000 m/s. To avoid this problem students should use the power of 10.

$$10^0 = 1$$
$$10^1 = 10$$
$$10^2 = 10 \times 10 = 100$$
$$10^3 = 10 \times 10 \times 10 = 1000$$
$$10^4 = 10 \times 10 \times 10 \times 10 = 10,000$$
$$10^5 = 10 \times 10 \times 10 \times 10 \times 10 = 100,000$$

The number of zeros represents to the power to which 10 is raised. It's called the exponent of 10. For example, the speed of light now can be expressed as:

$$3 \times 10^8 \text{ m/s}$$

In the following method, the number of places the decimal point is placed is equal to the exponential digit.

$$10^{-1} = \frac{1}{10} = 0.1$$

$$10^{-2} = \frac{1}{10 \times 10} = 0.01$$

$$10^{-3} = \frac{1}{10 \times 10 \times 10} = 0.001$$

$$10^{-4} = \frac{1}{10 \times 10 \times 10 \times 10} = 0.0001$$

$$10^{-5} = \frac{1}{10 \times 10 \times 10 \times 10 \times 10} = 0.00001$$

For example 0.0000332 is 3.32 x 10^{-5}, the decimal point is placed to the left which is equals to the value of the negative exponent.

The following general rule is useful:

$$10^n \times 10^m = 10^{n+m}$$

When dividing numbers expressed in scientific notation:

$$\frac{10^n}{10^m} = 10^n \times 10^m = 10^{n-m}$$

Basic algebra

The laws of arithmetic apply when basic algebra operations are performed. Symbols, such as x, y, and z are used to represent quantities that are not known.

Consider the following:

8x = 32
Solve for x,

x = 4   ;we get the answer by dividing both sides by 8

Next consider the equation

x+2=8

We subtract 2 from each side

x+2-2=8-2

x=6

In general, x + a=b, then x = b - a

Consider the next equation

$$\frac{x}{5} = 9$$

If we multiply each side by 5, we are left with x

$$\frac{x}{5}(5) = 9 \cdot 5$$

x = 45

All these cases, whatever operation we performed on the left side must also be performed on the right side.

The following rules apply:

Multiply: $\dfrac{a}{b} x \dfrac{c}{d} = \dfrac{ac}{bd}$ Divide: $\dfrac{(a/b)}{(c/d)} = \dfrac{ad}{bc}$

Adding: $\dfrac{a}{b} \pm \dfrac{c}{d} = \dfrac{ad \pm bc}{bd}$

Practice problems:

solve for x:

1. $a = \dfrac{1}{1+x}$

2. $4x-5=15$

3. $\dfrac{5}{2x+3} = \dfrac{4}{3x+1}$

Factoring

Useful formulas for factoring an equation:

$$ax+ay+az=a(x+y+z)$$ common factor

$$a^2+2ab+b^2=(a+b)^2$$ perfect square

$$a^2-b^2=(a+b)(a-b)$$ differences of squares

Linear equations

A linear equation is an equation of the form

$$y=mx+b$$

Where m and b are constants. Such equation is said to be linear because if we graph this equation, the graph of y versus x will be a straight line. The constant b, called the y intercept, is the value of y at x=0, represents the value at y at which the straight line intercepts the y axis.

The constant m is the slope of the line, and also equals to the tangent of the angle corresponding to change in x. The slope of the straight line can be expressed as

$$slope = \frac{y_2 - y_1}{x_2 - x_1} = \frac{\Delta y}{\Delta x} = \tan \theta$$

Note that if m < 0, the straight line has a negative slove, and at m > 0 has a positive slope.

There are no unique values of x and y, if x and y are both unknown in the equation.

Practice problems:

1. xy = 4 is a linear equation, is it true or false?

2. Draw the graph of the following straight line

$$y = 4x + 3$$

Quadratic equations

When a quadratic equation can't be solved by factoring, then we use a quadratic formula.

The general form of a quadratic equation is:

$$ax^2 + bx + c = 0$$

x is the unknown, a, b, c are referred as numerical coefficients of the equation.

The equation has two roots, given by

$$x = \frac{-b \pm \sqrt{b^2 - 4ac}}{2a}$$

if $b^2 \geq 4ac$, then the root is are real number

Practice problems:

1. $x^2 + 3x - 1 = 0$
2. $2x^2 + 4x - 9 = 0$
3. $2x^2 + 5x - 2 = 0$

Logarithms

$$x = a^y$$

if x is related to y, then the number a is called the base number, then the number y is the logarithm of x to the base a. thus,

$$x = \log_a y$$

Logarithms are exponents, listed below are some rules that will help you simplify terms.

if $y_1 = a^n$, and $y_2 = a^m$, then

$$y_1 y_2 = a^n a^m = a^{n+m}$$

Correspond to

$$\log_a y_1 y_2 = \log_a a^{n+m} = n + m = \log_a a^n + \log_a a^m = \log_a y_1 + \log_a y_2$$

$$\log_a y^n = n\log_a y$$

The two bases most often used are base 10 called common logarightsm, and logarithms to base e(where e=2.718...)called natural logarithm base.

$$y = \log_{10} x$$

when natural logarithms are used

$$y = \ln_e x$$

some useful properties of logarithms are

$$\log(ab) = \log a + \log b$$
$$\log(a/b) = \log a - \log b$$
$$\log(a^n) = n\log a$$
$$\ln e = 1$$
$$\ln e^a = a$$
$$\ln\left(\frac{1}{a}\right) = -\ln a$$

Geometry

In physics the basic analytical tools are geometric figures. Circles, and spheres are essential for understanding angular momentum and the probability densities of quantum mechanics. The ratio of the circumference of a circle to its diameter, d, is π, which has a value of 3.141592

Circle: $A = \pi r^2$ 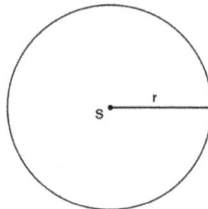 $C = \pi d = 2\pi r$

area of a circle   circumference of a circle

Parallelogram:
The area of a parallelogram is the base b times the height h.

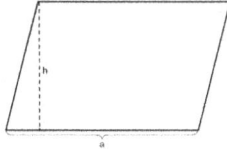

$$A = bh$$

Sphere:
A sphere of radius r has a surface area given by

$$A = 4\pi r^2$$

surface area of sphere

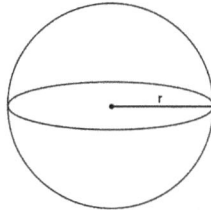

$$V = \frac{4}{3}\pi r^3$$

volume of a sphere

Cylinder:
A cylinder of radius r and length h has a surface are of

$$A = 2\pi rh$$

surface area of cylinder

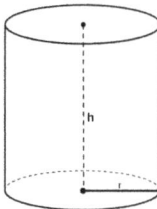

$$V = 2\pi r^2 h$$

volume of cylinder

Practice problems:

1. What is the area of a cylinder that has a radius that is ¼ its length?
2. What is the volume of a sphere, and cylinder if radius is 4.5?
3. Calculate the area of a cylinder, r=2.5, and h=7

Trigonometry

In physics, we will study vectors, and analyze motion in two dimensions. Trigonometric functions are also essential in analyzing period behavior, circular motion, and wave mechanics.

Triangle:

By definition, a right triangle is one containing at 90 degrees angle. The three basic trigonometric functions are defined by a triangle, such as sin, cos, and tangent functions. In terms of angle these fuctions are defined by

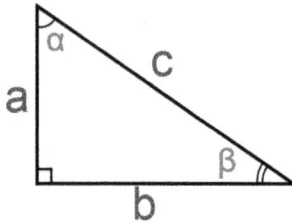

$$\sin \beta = \frac{a}{c}$$

$$\cos \beta = \frac{b}{c}$$

$$\tan \beta = \frac{a}{b}$$

The pythagorean theorem provides a relationship between sides of a right triangle:

$$c^2 = a^2 + b^2$$

it follows that

$$\sin^2 \beta + \cos^2 \beta = 1$$

$$\tan \beta = \frac{\sin \beta}{\cos \beta}$$

The functions cosecant, secant, and cotangent are defined by

$$\csc \beta = \frac{1}{\sin \beta}$$

$$\sec \beta = \frac{1}{\cos \beta}$$

$$\cot \beta = \frac{1}{\tan \beta}$$

Practice problems:

1. If a=3, b=4, c=5 from triangle find (a) cos (b) sin (c) tan
2. If a=5, b=4 then use pythagorean theory find c
3. From problem 1 find (a) csc (b) sec (c) cot

CHAPTER 5

PHYSICS AND MEASUREMENTS

Physics is based on experimental observations and quantitative measurements. The laws of physics are expressed in terms of physical quantities that required a clear definition. Physical quantities are numbers that are obtained by measuring the physical world; physical quantities are force, velocity, volume, and acceleration that can be described in terms of measurements.

For example, the length of this book is a physical quantity, as is the amount of time it takes for you to read this section, and the temperature of the air in your room.

A measurement of any physical quantity involves assigning that quantity to a defined standard unit. In mechanics, the three basic quantities are length(L), mass(M), and time(T). Results in each physics problem must be reported in units. For example, to measure the distance between two points, we need a standard unit of distance, such as an inch, a meter, or a kilometer. If the result of a problem is 15 meters, it means that it is 15 times the length of the unit meter. If you write the result as 15, it has no meaning, your professor may assume the answer is in seconds, pounds, or feet. You can ask yourself "why is it necessary to know this?". In physics many of the quantities that you will be studying are: velocity, force, momentum, work, energy, and power that can be expressed in terms of time, length, and mass.

In 1960, an international committee established a set of standards for the scientific community called SI(French Systeme international).

There are seven base quantities in the SI system.

Length
Mass
Time
Electric current

Thermodynamic
Temperature
Amount of substance
Luminous intensity

I will define three base SI units length, mass, and time. Later, when you study thermodynamics, and electricity you will need to know the other base SI units.

Length:

The meter(m) is the SI unit of length.

In 1960, the length of the meter was defined as the distance between two lines on a specific platinum-iridium bar stored under controlled conditions. Until recently, the meter was defined as 1,650,7653.73 wavelengths of orange-red light emitted from a krypton-86 lamp. However, the latest definition, the meter is determined using the speed of light through empty space, which is defined exactly 299,792,458 m/s. The meter is then, the distance light travels through a vacuum during a time of 1 / 299,792,458 second.

In the British engineering system, the unit of length is, the foot (ft). The SI unit system is universally accepted in science and industry. Make sure to differentiate between the two systems when you are solving physics problems.

Mass:

The basic SI unit of mass, the kilogram(kg).

The kilogram is now defined to be the mass of a specific platinum-iridium alloy cylinder. Another unit, which is a U.S customary system, of force(-pound-force) rather than mass is considered a base unit.

Time:

The basic SI unit of time, the second, is defined as periods of the radiation from a cesium-133 atom exactly as 9,192,631,770 cycles per second.

The unit of time, second(s), was defined in terms of the rotation of the earth. The mean solar day is equals to (1/60)(1/60)(1/24). Currently, the second is now defined in terms of the frequency of a particular kind of cesium atom

Conversion of units

Because different systems of units are in use, sometimes it is necessary to convert units from one system to another. Units of length are as follows:

1 mile = 1609 m = 1.609 km

1 ft = 0.3048 cm = 30.48 cm

1 m = 39.37 inches = 3.281 feet

1 in. = 0.0254m = 2.54 cm

Learn to convert between SI unit system, and the British engineering unit system. You will be saving time if you master this before jumping in to solve problems.

For example, suppose we wish to convert 10 inch to centimeters. We know 1 inch = 2.54 cm.

We find that

10 in. = 10 in. (2.54 cm / 1 in.) = 38.1 cm

10 inches equals to 38.1 centimeters.

CHAPTER 6

HOW TO WORK PHYSICS PROBLEMS

The theory of problems

Every single word problem has three components, and a problem solver has to dissect the problem, recognize these three components, find what information is irrelevant or relevant, and provide a solution.

The three components are information concerning given expressions, operation expressions, and goal expressions. A problem solver defines each expression according to the material he learned in the chapter. Generally, a student may know all the formulas, definitions, so on, but may not know how to associate a definition to the problem, and may not know how to differentiate what is relevant, and irrelevant information. However, in some cases, the problem will be obvious that a particular problem may ask for a mathematical operation.

Given expression:

The information that is clearly stated in the problem is the given. It can be a constant number, a mathematical equation or can be a principle. These are the known values explicitly stated in the problem. For example, the mass of an object is 4 kg, initial speed is zero, the force and mass are constant, etc. Reduce the problem to its essentials, list the known quantities, and the required quantities. Draw and label a diagram if possible.

Ask yourself this question:

Is the information given relevant or irrelevant to the problem?

Operation expression:

An operation expression are the steps taken to find the unknown, and the principles related to a problem are identified. The known values may give you an idea on what formula to use to solve a problem.

The operation expressions are steps to find everything you need before performing a calculation on a problem. The thinking process is done at this stage. Analyze the problem, think about it, and convert units if you can.

A problem solver has to ask himself these questions in an operation.

- What are the known and unknown values?
- What concepts and principles are involved?
- How do known values relate to a formula(equation)?

Goal expressions:

The goal expressions are the final steps taken to solve a problem. The goal of the problem. At this stage you want to know what is the problem asking for.

Read the question of the problem carefully, and perform any calculations. Solve algebraically as much as possible, and complete the numerical solution.

Questions to ask yourself:

- What is the goal of the problem?

- Do I need to do a calculation or explain a principle?

- What is the problem asking for to solve?

- Do the units match?

- Is the answer reasonable?

You should pay careful attention to physics problems and should not be discouraged if you don't perfectly understand the concepts. If you can't understand a problem, ask your class friend for a hint or contact your instructor.

Physics problems

1. You decide to go to a long trip. You drive for one hour at five miles per hour. Then two hours at four miles per hour, and then three at seven miles per hour. How many miles did you drive?

Given: One hour at 5 mph
        Two hours at 4 mph
        Three hours at 7pmh.

The problem is asking for distance. To get the answer we use equation:

$$v = \frac{x}{t} \qquad\qquad x = vt$$

x - distance

t - time

v - speed

While you drive the speed changes, so we have to split the trip into three sections.

Operation:

$$X_1 = 1 \times 5 = 5 \text{ miles}$$

$$X_2 = 2 \times 4 = 8 \text{ miles}$$

$$X_3 = 3 \times 7 = 21 \text{ miles}$$

Xtotal = X_1 + X_2 + X_3

Xtotal = 5 + 8 + 21 = 34 miles

Goal: You drove 34 miles in total.

2. Visualize the problem

A stone is thrown from the top of a building straight upward, as it reaches the maximum height its velocity is zero at an instant of a second. What is its acceleration at this point?

In this type of problem there is nothing to calculate. I want you to visualize the problem in your mind.

The velocity is zero at an instant of time, but still changing in velocity. The change of velocity slows down for a second. Picture the stone frame pauses for a second as it reaches its maximum height. Since the stone will certainly fall it had an acceleration before it reached a velocity of zero, and it will have an acceleration after it pauses for a few seconds.

Newton's second law states:

The acceleration of an object is directly proportional to the net force acting on it, and inversely proportional to its mass.

Gravity will act on the stone at all points on its path, producing a net force, and also producing a constant acceleration at all points in its path.

The velocity is zero, but still has a rate of change of velocity, in this case, it is the gravitational force. The acceleration at that point is 32 ft/s^2

3. Conceptial problem

A roman soldier carries a heavy shield to protect himself against iron arrows. Does the shield protects him from the arrows' momentum or kinetic energy?

An enemy arrow has momentum, and kinetic energy, and when it strikes the roman soldier's shield it will pushed the roman soldier back a bit. The shield protects the roman soldier from kinetic energy, not momentum.

HOW TO STUDY PHYSICS

4. Problem strategy

How to apply Newton's second law properly on problems.

Newton's second law states:

The acceleration of an object is directly proportional to the net force acting on it, and inversely proportional to its mass. To simplify it the acceleration is dependent on two variables: the force acting on an object, and the mass of the object.

First step identify all the forces acting on the object. Pay close attention to the direction of the acceleration vector of an object. If you know the direction of the acceleration an arrow will help you choose the best coordinate axes.

1. Draw a diagram with labels

2. Identify each force acting on each object(if more than one)

3. If the direction of the acceleration is given, then choose a coordinate axis parallel to that direction. For example if an object is sliding down, show a vector showing gravitational force acting on it, and another force parallel to the surface.

4. Apply newton's second law.

5. Solve the problem.

Check for the correct units.

5. Problem strategy

How to solve motion along a curved path problems.

To understand the problem label the force as tension force, or normal force. Draw a diagram of the object. Include coordinate axes, and show the point of origin. Draw one coordinate axis in the direction of motion, and second in the centripetal direction.

Apply Newton's second law in componet form. Take into account net forces acting on object.

If possible substitute acceleration with v^2 / r, v is the speed, and r is the radius.

Lets assume the object moves in a circle of radius r with a constant speed, then use $2(3.14)r/T$, in which T is the time for one revolution.

6. You suspend a bowling ball beneath the surface of the water with a string. When is fully submerged under water you find it feels less heavy. As the bowling ball is submerged still deeper, is the force needed to hold the bowling ball the same?

Yes, The buoyant force doesn't change with the depth of the water, so the force will be the same.

7. Problem strategy

How to solve simple harmonic motion problems

For a spring: if the spring is extended then choose positive x direction.

1. Use the equation for simple harmonic motion

$$F_x = -kx$$

2. Don't use the kinematic equation for constant acceleration.

3. When calculating trigonometric functions make sure the calculator is in the appropriate mode(degrees or radians)

8. A block of mass 1.8 kg is attached to a horizontal spring that has a force constant of 1.2 x 10³ N/m. The pring is compressed at a distance of 3.0 cm, then the block is released from rest. see figure.
Calculate the speed of the block at x = 0.0 cm as it is released

First lets consider the surface is frictionless. In this problem we need to pay close attention to units. Notice the distance is given in cm, but we need it in meters. We need to convert.

Figure

Given:

mass of block: 1.8 kg
k = coefficient = 1.2 x 10³ N/m
Distance spring is compressed: 3.0 cm

The principle is work-energy theorem which states the work done by a constant net force Wnet in moving an object is equals the change of kinetic energy of the object.

Wnet = Kf - Ki

We will first calculate the work done by the spring, Wspring, then we will use this number to find the velocity with the following equation :

$$W_{spring} = \frac{1}{2} k x_m^2 \qquad W_{spring} = \frac{1}{2} m v_f^2 - \frac{1}{2} m v_i^2$$

Operation:

Convert distance of the spring:

-3.0 cm = -3.0 x 10^{-2} m   ; negative value since the block is pushed

Use Wspring equation to calculate the work done by the spring.

$$W_{spring} = \frac{1}{2} k x_m^2$$

Wspring = 1/2(1.2 x 10^3 N/m)(-3.0 x 10^{-2} m)^2= 0.54 J

Using work-energy theorem gives:

$$W_{spring} = \frac{1}{2} m v_f^2 - \frac{1}{2} m v_i^2$$

Goal:

Wspring = 1/2(1.8 kg)V_f^2 - 0

Note: We need to calculate the final velocity, given initial velocity at zero.

0.54 J = 1/2(1.8 kg)V_f^2

V_f^2 = 1.08 / 1.8 kg = 0.6 m^2/s^2

V_f = 0.77 m/s

9. A ring is heated until it expands one percent. Will the diameter of the hole increase or decrease?

This physics problem doesnt require calculation. Metals expand with heat. The ring diameter will increase as you apply heat.

10. Problem strategy

How to solve friction problems.

There are four different types of frictions: static, kinetic, sliding, and fluid. **Static friction** takes place when there is no rolling involved between the object, and the surface. The force of static friction pushes the object back opposing it. In general, static friction is the force that keeps the box from slidng. **Rolling friction** is the force that opposes the rotation of a motion. For example a wheel rolls at a constant speed on a horizontal surface has no rolling friction, as there is no force opposing it. **kinetic friction** is a force opposing the motion of an object. It is a force that doesn't depend on the applied horizontal force, but depends on what materials the surface is made of, and temperature of the surface. **Fluid force** is the opposing force of liquids or gases.

Steps:

1. Draw a diagram with y and x axis, with x-axis parallel to the contacting surface. Draw an arrow signifying the direction of the frictional force in a way that is opposes the sliding or motion force.

2. Sum all the forces, and solve for the normal Force.
   If there is kinetic force relate normal and frictional force with the coefficient.

3. Apply total forcce to the object, and solve for the quantity.

11. Problem Strategy

How to solve problems relating to collisions betwen two objects

Steps:

1. Label objects, setup a coordinate system and define the velocities.

2. In the coordinate system draw an arrow indicating the directions, before or after impact. Add the velocities values to have an idea on the speed of each item.

3. Include signs for velocity vectors, write an equation for the x and y coordinate system for the momentum of each object before or after impact.

4. Solve.

12. A 4.0 kg brick is attached to a spring. It oscillates with an amplitude of 5.0 cm. It has a period of 2.0 s.

a. What is the total energy?

Given:

Amplitude: 5.0 cm
Period(T): 2.0 s

Energy: $E = \frac{1}{2}kA^2$

Force constant: $k = m\omega^2$

Angular frequency: $\omega = \frac{2\pi}{T}$

Operation:

Convert Amplitude from centimeters to meters:

5.0 cm = 0.050 m

Substitute force constant, and angular frequency into energy equation

$$E = \frac{1}{2}m\left(\frac{2\pi}{T}\right)^2 A^2$$

$$E = \frac{1}{2}(4.0\,kg)\left(\frac{2\pi}{2.0s}\right)^2 (0.050m)^2$$

$$4.93x10^{-2}\,J$$

Goal:

The total energy is: $4.93x10^{-2}\,J$

CHAPTER 7

PHYSICS IS SO COMPLICATED

If you are having trouble with a course, ask yourself two questions: what are you doing wrong?, and why you think you are having difficulty? Be as honest as possible, try to find a solution, try to do things a little different, and try to learn your own way. You will find physics difficult if you don't find it interesting, otherwise if you like physics, want to learn it, and you are willing to study, then you will succeed.

These are some solutions:

- If the textbook doesn't explain the material clearly, then try going to your college or local library to find different Physics book, in which you will find some books simpler and easier to understand than your own book.

- If you plan to keep your own book, then highlight important points, and write down your own comments in the margin.

- Talk to your instructor, he may suggest a tutor, sometimes a good tutor can teach you better than your instructor.

- Pay special attention to definitions. Use a dictionary to find clear and concise meaning of words.

- Before taking an exam, revise your lecture notes, re-read the chapter, check yourself whether you can solve a physics problem without looking at an example. It is essential and most important to learn to solve all problems shown in class by your instructor.

- Seek the advice of a student who has taken the course.

- Mathematics is one of the most important tools for engineers, and scientist, the more math you know, the better you can solve problems. You may to need to brush up on your math. Read chapter mathematics in physics, and try to solve practice problems.

Problem solving strategies and hints are included throughout the textbook to help you solve problems. Footnotes are sometimes used to supplement an example or to cite a reference to a concept, make sure to read all footnotes. You may discover later on in class that you didn't know the material as well as you thought you did, and is better to find out while studying than during the exam. Review every day your notes, and on weekends solve problems.

Techniques to study

You have to realize that the reason why it is necessary to study is so that you learn the material. If you delay or pospone to learn the material; it won't be easy to learn most of it in a short time. Some students can perform better under pressure, and need to worry about a grade in order to start to study. In some cases, it is not easy to persuade oneself to start studying. If you procrastinate to start studying these are some methods that might work for you.

Time

Most people who procrastinate generally don't have a sense of time, that time is on their side, and that they have control of it. Most feel that they will have enough time to complete their homework or to study later on during the day.

Intelligence

Most people who delay to study feel that they are very smart who have the ability to remember, and apply all concepts and principles. Most feel that they understand the material because the instructor explained everything clearly. There is no need to devote more time to what is explained.

Methods to study

1. If you are browsing the internet, and can't persuade yourself to start studying. Open your physics book to any chapter, leave the book open, and flip the pages until you find an interest or the need to study.

2. It is difficult to get into a mood to start studying, in some cases listening to music helps, play a movie, and let it play in the background as you concentrate on the assignment. Hyperactive minds tend to do multiple things at the same time, and seem to do just fine on learning.

3. Think of the time you have left to turn in an assignment, quiz, midterm exam or final exam. Probably you have one week, a weekend, or 1 day. If you consider in a 24 hours period. You will need 8 hours of sleep, 8 hours to work, 3 hours to eat, 1 hour to do errands, 1 hour to do housework, 1 hour to talk to your friends, 1 hour for church, 1 hour to drive from point A to point B, 1 hour to buy the house food, 1 hour to read the news etc. If you take all of these things into account you will find out that you don't really have enough time and generally you will have little time to study, if you find time mostly at night. Every single day remind yourself why it is necessary to study, because you won't have enough time to study later on. Say this to yourself, "I need to study now or I won't have enough time to study later on".

4. Stop whatever you are doing and think for a second. Consider that time moves fast, seconds quickly turn into minutes, minutes quickly turn into hours, and that the day will have a morning, an afternoon, and before you notice it will be evening. Be aware that time moves fast, if you don't find the reason to study now then the day will end, and you won't have enough time to study the following day.

5. Consider where you are now, in reference to time, think for a moment what you want to accomplish. Ask yourself how are you going to get there, and what you need to do. Your answer will lead to "I need to learn, and to learn I need to study". Every person in order to become someone

in life needs to study. You need to think of yourself, you can't continue to procrastinate anymore, and you have to take action now.

CHAPTER 8

EXPERIMENTS IN LABORATORY

Physics laws and theories are tested in the laboratory. It is a science based upon experimental observations, and can be the most exciting part of the course. The purpose of laboratory work is to test ideas, and models discussed in class or textbook. In the experiments you will see physics in action; it will help you understand and remember what you discussed in class. It gives you an opportunity to practice laws, some skills in the use of scientific instruments and techniques. In your first year of physics you will most likely explorer physics, learn how to do measurements, and discover new principles. It is true that most likely that you will not make a great discovery; the first year will be simply a learning adventure.

Before going to a physics laboratory, study the laboratory manual so that you will know what you are going to do, and you can plan in advance how to use your time efficiently. Make an effort to learn as much as you can, and while doing experiments take multiple readings to get an idea of the reliability of your measurements. Pay attention to the derivations and the equations used, while doing calculations, when you substitute values into the equation you will know why you use them.

Have fun while in laboratory, keep your mind open, and alert, if possible try things out not asked in the instructions to put your own ideas to the test.

Always take your physics book to the laboratory. Take a notebook and a pen to class to take notes on your observations.

Mechanic experiments

You will study friction, force, momentum, law of gravity, energy, velocity, vectors, equilibrium etc. The purpose of these experiments is to direct your thinking as to how physics principles work. The centrifugal force experiment you will learn about inward force, and how to calculate the force in an unbalanced tire traveling at high speed.

Electricity experiments

You will measure current in a circuit, you will calculate the resistance, and then measure ohms with a volt meter. In more advance experiments you will build a circuit diagram.

Optic experiments

You will learn to calculate the focal length. You will compare focal length with experimental value found in class, and with the calculated value using formulas.

Ask yourself these questions:

- What is the purpose of this experiment?

- Why do we do it this way?

- What does this measurement show or prove?

- Do I have all the measurements I need to complete my experiment?

Writing laboratory reports

The goal of every single experiment is to make the student realize that the laboratory work has applications outside the laboratory. A student has to write a report to show the instructor he understands how physics work. Writing a scientific report is the most important tool of every scientist and engineer. The laboratory report teaches a student to put his thoughts in writing, because it takes mind power to transfer thoughts into paper. The report should be clear, its purpose is to convey information to the readers rather than to puzzle them. Every scientist needs to learn the ability to express himself clearly. Be precise, and concise in writing laboratory experiments. When writing a laboratory report explain the experiments with simple words, don't try to impress the instructor, and should be easy to read.

CHAPTER 9

HOW TO TAKE EXAMINATIONS

If you have studied carefully, and really know what you have studied, then you are most likely ready for the exam. If you have been studying everyday then you have nothing to worry about. Remember there is no technique to taking examinations, and systematic instructions which you may follow. On the day of the exam get up early so you can take a long shower, take your time eating a healthy breakfast, and drive or walk at the right pace to the exam. Arrive early if possible, sit calmly, get your pencil, paper, calculator, glasses(if you need them), notes and eraser ready.

Taking an exam

1. Read the questions carefully. Write your name, and date on the exam. Find out exactly what each problem is about. Check on the number of questions to be answered. Quickly browse through the exam to know what to expect. Don't rush; work at a convenient pace but without wasting time.

2. Answer first the questions on which you know the most. If you think you might run out of time be sure you have answered the easy ones. Don't waste time on the hard problems. Don't worry about how you are doing. Don't spend too long on any one question.

3. Pay no attention to others. Budget your time.

4. Write down full required work on the answer sheet in orderly, clean, and label your answers, if possible. Don't erase anything if you have doubts,

don't leave any questions blank, you might get partial credit for trying or might be right after all.

5.   Check your work, and remember to put down the units. The ability to follow instructions counts a lot in an examination. Don't second guess your answers if you have any doubts.

6.   Ten minutes before the examination is over, take about one minute to check your work to make sure you have made no major mistakes. Don't leave any questions unanswered; write down the best you know.

Ask yourself these questions before submitting your exam

- Have I fully answered all the questions?

- Do I need to convert units?

- Do I feel good about the answers?

- Am I forgetting to do anything?

- Am I ready to submit the exam?

Most instructors are not planning to fail students; they actually want students to do well on exams. Instructors prepare exams so that ordinary and careful students can pass them. After the exam papers have been returned to you, be sure to review the points you missed. If you were graded poorly, then you need to sit down and figure out what you did wrong. Learn from your mistakes, and promise yourself you will do better on the next exam. If you did great, good job, you have taught yourself how to study physics, developed an organized system for examining problems, and extracting relevant information. You will become a more confident problem solver.

This page intentionally left blank

This page intentionally left blank

HOW TO STUDY PHYSICS

APPENDIX

Constants

Acceleration of gravity at Earth's surface	$9.81 \text{ m/s}^2 = 32 \text{ ft/s}^2$
Speed of light	$2.997,924.58 \times 10^8 \text{ m/s}$
Avogadro's number	$6.022\ 1415 \times 10^{23}$ particles/mol
Speed of sound	331 m/s
Density of air	1.217 kg/m^3
Escape speed	$1.12 \times 10^4 \text{ m/s}; 6.95 \text{ mi/s}$
Mass of Earth	$5.97 \times 10^{24} \text{ kg}$
Radius of Earth	$6.37 \times 10^6 \text{ m}; 3960 \text{ mi}$
Solar constant	1.37 kW/m^2
Standard temperature	$273.15 \text{ K } (0.00°\text{C})$
Standard pressure	$101.325 \text{ kPa } (1.00 \text{ atm})$
Fundamental charge	$1.602\ 176\ 453 \times 10^{-19} \text{ C}$
Mass of electron	$9.109\ 382\ 6 \times 10^{-31} \text{ kg}$
Planck's constant	$6.626\ 0693 \times 10^{-34} \text{ J s}$
Mass of proton	$1.672\ 621\ 71 \times 10^{-27} \text{ kg}$

Conversion Factors

Length

1 m = 39.37 in = 3.281 ft = 1.094 yd
1 m = 10^{15} fm = 10^{10} Å = 10^9 nm
1 km = 0.6214 mi
1 mi =5280 ft = 1.609 km
1 lightyear = 9.461 x 10^{15} m
1 in = 2.540 cm

Time

1 hour = 3600 seconds
1 year = 365.24 days = 3.156 x 10^7 s

Mass

1 tonne = 10^3 kg = 1Mg
1 slug = 14.59 kg
1 kg = 2.21 lb

Volume

1 Liter = 1000 cm^3 = 10^{-3} m3
1 gal = 3.785 L

SI Units

Length	Meter	m
Mass	Kilogram	kg
Time	Second	s
Electric current	Ampere	A
Temperature	Kelvin	K
Amount of substance	Mole	mol
Luminous intensity	Candela	cd

This page intentionally left blank

This page intentionally left blank

BIBLIOGRAPHY

Simon, H. A.,& Newell, A. Human problem solving, American Psychologist, 1971

Polya, G., How to solve it, Garden City, N.Y.: Doubleday & Company, 1957

Wickelsgren, Wayne, How to Solve Problems, San Francisco, CA,: W. H. Freeman and Company, 1938

Serway, Raymond A., Physics: for Scientists & Engineers, 4th Edition, James Madison University, Saunders College Publishing, 1996, 1990, 1986, 1982

Leighton, Walter, A First Course in Ordinary Differential Equations, University of Missouri, Wadsworth Publishing Company, Inc. Belmont, CA, 1976

Snell, T. Cornelia, Snell, Foster Dee, Ph. D., Chemistry Made Easy: Volume one, The theory of inorganic chemistry, Van Nostrad Company, NY, 1943

Belvoir, Fort.,How to study and take examinations, US Army Engineer Training Brigade, US Army Engineer School, 1984

Chapman, Seville., How to Study Physics, Addison-wesley press, Cambridge, Mass., 1949

Mann, Charles Riborg, The teaching of physics for purposes of general education, The Macmillan company, NY., 1917

Chessin, P. L., Problem for solution: American Mathematical Monthly, 1954

Giancoli, Douglas C., Physics: Principles with Applications, fifth edition, Prentice Hall, Upper Saddle River, New Jersey 07458, 1998, 1995, 1991, 1985, 1980

HOW TO STUDY PHYSICS

www.ingramcontent.com/pod-product-compliance
Lightning Source LLC
Chambersburg PA
CBHW070946210326
41520CB00021B/7081